Nonlinear Network's Neural Nexus

Exploring Complexity, Connectivity, and Emergent Phenomena in Network Science

Benjamin Langley

Table of Contents

1 Introduction to Networks

Networks, at their simplest, are collections of interconnected entities. These entities, which we call nodes or vertices, represent anything from individuals in a social network to computers on the internet or neurons in the brain. The connections between these nodes, known as edges or links, define the relationships and interactions. Understanding this fundamental structure is crucial to appreciating the intricacies of network science. It's the foundation upon which we build our understanding of complex systems.

A helpful way to visualize these networks is through graph theory. Think of it as the mathematical language of networks. Graphs, with their nodes and edges, provide a framework for representing and analyzing network properties. We can use graph theory to quantify concepts like the degree of a node (the number of connections it has), the shortest path between two nodes, and the overall connectedness of the network. These metrics are invaluable tools for understanding network structure and function. For instance, a highly connected node might represent an influential person in a social network or a critical component in a technological infrastructure.

Beyond simple visualizations, graph theory enables more sophisticated analyses. We can explore the distribution of node degrees, revealing whether a network exhibits a random structure, a scale-free structure (with a few highly connected hubs), or something in between. These structural characteristics profoundly impact the way information flows and how resilient the network is to disruptions. A scale-free network, for instance, is often more robust to random failures than a random network, because it lacks a single point of failure. The study of these properties is paramount in understanding and predicting the behavior of various real-world systems.

Consider the implications. Imagine analyzing a social network to identify key influencers, or studying a biological network to understand disease propagation. By applying graph theoretical tools, we can extract meaningful insights from seemingly complex datasets. These methods allow us to quantify and understand the patterns and relationships that govern the behavior of networks, revealing hidden connections and unforeseen consequences. This provides a significant advantage, enabling researchers to anticipate outcomes and create targeted interventions.

Path lengths, another key concept in graph theory, tell us how many steps it takes to navigate between any two nodes. A small-world network, characterized by short average path lengths, is particularly efficient for information transmission. This explains why information spreads rapidly in networks like social media. Analyzing these path lengths provides a deeper understanding of the ease with which information or influence travels within a particular network, enabling us to build more efficient communication and influence strategies. For instance, understanding the structural constraints on the path lengths enables targeting communication for maximum impact.

Centrality measures, which identify the most important nodes within a network, provide yet another crucial analytical lens. Different centrality measures exist, each focusing on a specific aspect of node importance–degree centrality focuses on the number of connections; betweenness centrality highlights nodes bridging distinct parts of the network; and closeness centrality emphasizes nodes with short paths to all other nodes. Understanding the centrality of certain nodes allows us to focus interventions on those that have the greatest impact. This approach is relevant whether focusing on mitigating the spread of disease or optimizing the functionality of a technological network. By identifying key nodes, we can then strategize how to control or improve network behavior effectively.

Furthermore, the analysis of networks extends beyond static structural properties. Network dynamics, examining how networks change over time, are just as important. The evolution of connections, the addition or removal of nodes, and the shifts in information flow are all critical aspects of real-world systems. The ability to model these dynamic processes allows for a more realistic and nuanced understanding of how networks behave and adapt. For example, the spread of a disease in a population isn't static; it evolves as individuals become infected and recover. Understanding this dynamism is essential for effective intervention strategies.

Understanding network basics, coupled with the mathematical tools offered by graph theory, lays the groundwork for more advanced explorations into nonlinear dynamics, neural networks, and emergent phenomena. The insights gained from these fundamental concepts provide a powerful lens through which to examine a diverse array of complex systems. The application of these principles is profound, impacting our understanding of everything from brain function to social interactions and technological infrastructure.

1.1 Network Basics

Networks, at their most fundamental level, are collections of interconnected entities. Think of a social network, where individuals are connected through friendships, or the internet, a vast web of computers communicating with each other. These entities, which we call nodes or vertices, are linked by relationships or connections we refer to as edges or links. Understanding this simple yet powerful framework is the cornerstone of network science. The relationships between nodes aren't arbitrary; they define the structure and, crucially, the function of the entire network. This structure dictates how information flows, how quickly changes propagate, and even the network's overall resilience to disruptions. The characteristics of these connections – their strength, directionality, and even their temporal dynamics – all contribute to the network's overall behavior, making it far from a simple sum of its parts.

A key concept in understanding network structure is the degree of a node. This refers to the number of connections a single node has. A highly connected node, with a large degree, might be considered influential or central within the network, while nodes with few connections might be more peripheral. The distribution of these degrees across the entire network–the degree distribution–provides crucial insights into its overall architecture. For example, a network with a power-law degree distribution, a common occurrence in many real-world networks, exhibits a few highly connected nodes (hubs) and many nodes with few connections, a characteristic quite unlike a random network where node degrees tend to be more evenly distributed.

Beyond simple degree, we need to consider the distances between nodes. The shortest path between two nodes, called the geodesic distance, tells us how efficiently information can travel. Networks with small average geodesic distances, also known as small-world networks, are characterized by their efficient information transmission capabilities. This concept explains the famous "six degrees of separation," suggesting that most people are connected by a chain of just six acquaintances. But not all networks are small-world networks. Some are characterized by long average distances, hindering efficient information flow. The existence of "hubs" and their positions in the network also directly impact this efficiency. A network's ability to withstand failures depends on the distribution of these connections and the ways in which nodes can reroute information if a major node or connection fails.

Network centrality measures provide further insights into the importance of individual nodes within the larger network. These measures quantify the influence or importance of a node based on its position and connections. Different centrality measures exist, capturing various aspects of node importance. Degree centrality simply counts the number of connections, while betweenness centrality measures the number of shortest paths that pass through a node, indicating its importance in information transmission. Eigenvector centrality considers both the number of connections and the centrality of connected nodes, reflecting influence within the network. Understanding these different centrality metrics helps us identify key nodes that, if removed or disrupted, could significantly impact the overall network functionality. We can also use these metrics to analyze the network structure, by plotting the centrality measures across different nodes, revealing information on how the information flows across these networks.

Beyond the static structure, many networks are dynamic, with edges constantly changing. This dynamism is crucial to consider for a complete understanding. Edges may appear and disappear, reflecting changing relationships or connections. This leads to temporal networks, where the network structure evolves over time, often in complex and unpredictable ways. Analyzing the dynamics of these networks adds another level of complexity, requiring techniques that can track changes in structure and connections and their effect on how the system behaves. The structure and properties of the network influence the way that information propagates throughout the system, as well as the system's overall stability and resilience. These considerations become even more complex when we account for the dynamics that may be present within the network. A simple change in the connections of a network can drastically alter its global properties, which can be extremely important to consider when we study how complex systems function.

1.2 Graph Theory Fundamentals

Graphs, at their core, are abstract representations of relationships. Think of them as blueprints, sketching out how various entities interact. These entities, known as nodes or vertices, represent anything from individuals in a social network to neurons in your brain or routers on the internet. The connections between them, the edges or links, signify the interactions or relationships, be it friendship, synaptic connection, or data transmission. Understanding this fundamental duality – nodes and edges – is the first crucial step. From this simple foundation, we can build a sophisticated understanding of complex systems.

The degree of a node, a seemingly simple concept, reveals much about its influence within the network. It simply counts the number of edges connected to that particular node. A highly connected node, with a large degree, often acts as a hub, a central point of communication or influence. Consider a social network; an individual with a high degree is likely very influential, easily spreading information or initiating trends. On the other hand, nodes with low degrees often remain peripheral, their influence restricted to a smaller circle. This seemingly simple metric provides a powerful insight into network structure and function.

Paths, sequences of connected edges, form the arteries of information flow within a graph. The shortest path between two nodes, determined by algorithms like Dijkstra's algorithm, dictates the efficiency of communication or the speed of signal transmission. In a social network, the shortest path between two individuals represents the most efficient route for information to travel. In a computer network, it determines the speed of data transmission. These seemingly simple paths hold the key to understanding efficiency and resilience within many different complex systems.

Centrality measures go beyond merely looking at individual node degrees. They provide a more holistic view, quantifying a node's importance within the entire network. Several measures exist, each revealing different aspects of influence. Degree centrality, as discussed, focuses solely on connections. Betweenness centrality considers a node's role in connecting other nodes. A node with high betweenness acts as a bridge, mediating connections between other groups. Closeness centrality measures how easily a node can reach all other nodes in the network. Understanding these different types of centrality allows us to identify key players and critical infrastructure within a complex network.

Beyond basic structural analysis, graph theory provides tools to characterize the overall network structure. The distribution of node degrees, for instance, isn't always uniform. Some networks exhibit a scale-free property, meaning a few nodes have an extraordinarily high degree, while most have relatively few connections. This creates a highly uneven landscape of influence. Other networks may exhibit small-world properties, characterized by short average path lengths and high clustering coefficients. These properties impact how information spreads and how resilient the network is to failures or attacks.

Understanding these characteristics is paramount for analyzing and managing various systems. For instance, recognizing the scale-free nature of the internet helps in designing robust routing protocols that can withstand cascading failures. Similarly, understanding the clustering in social networks facilitates targeted advertising or the prediction of information diffusion patterns. The mathematical tools of graph theory, therefore, are indispensable for navigating the complexities of interconnected systems. They move beyond simple descriptions and provide predictive capabilities, enabling a deeper, more quantitative understanding. This understanding allows for a more effective management and control over these complex systems.

This chapter only scratches the surface of graph theory's power. Future chapters will delve deeper into more advanced concepts and build upon these fundamentals. However, mastering these core ideas–nodes, edges, paths, centrality, and overall structural properties–is fundamental to navigating the intricate world of nonlinear networks and their emergent behaviors. Remember that graphs are more than just diagrams; they are mathematical models that can unlock profound insights into a wide range of complex systems. These core principles are vital for gaining a profound understanding of the networks discussed within this book. This chapter has laid the groundwork for understanding the structure of the networks that we will encounter later in this text.

2 Nonlinear Dynamics in Networks

Imagine a network, not as a static structure, but as a vibrant, ever-shifting landscape. Nonlinear dynamics inject this dynamism, transforming predictable behavior into a realm of unpredictable complexity. We're not talking about simple cause and effect; instead, tiny initial differences can blossom into dramatically divergent outcomes, a phenomenon often described as the butterfly effect, where a butterfly flapping its wings in Brazil could theoretically cause a tornado in Texas. This inherent unpredictability challenges traditional linear models, demanding a shift in perspective toward the richness of nonlinearity.

Nonlinearity isn't merely a quirk; it's a fundamental feature shaping network behavior. Consider a network of neurons. The firing of a single neuron isn't merely an isolated event; it triggers cascades of activity, influencing the firing patterns of connected neurons. These interactions, often nonlinear in nature, lead to emergent patterns that are impossible to predict simply by studying individual neurons in isolation. The overall network behavior transcends the sum of its parts. This holistic perspective underscores the need to consider the collective dynamics, rather than focusing exclusively on individual components.

Bifurcations represent pivotal moments in a network's evolution. Imagine a river's gentle flow suddenly transforming into a raging torrent due to a shift in the landscape. Similarly, a network might experience a sudden transition from a stable state to a chaotic one. Such transitions are often abrupt, marking significant shifts in the network's overall behavior. Identifying these bifurcation points is crucial; it allows us to understand and predict critical transitions that could be beneficial or detrimental depending on the specific network. These transitions might enhance information processing or lead to network collapse.

Synchronization, a captivating manifestation of nonlinearity, occurs when independent network components begin to oscillate in unison. Think of fireflies blinking in synchrony, or a group of musicians playing together in perfect harmony. In networks, this synchronization can arise spontaneously, leading to coordinated activity that would be impossible without nonlinear interactions. Such synchrony can enhance the efficiency of information processing, but it can also make the network vulnerable to disruptions; if one element fails, the entire synchronized system could be affected.

The implications of nonlinear dynamics extend far beyond theoretical considerations. Understanding these dynamics is essential for designing robust and efficient networks. Imagine designing a communication network that can withstand unexpected surges in traffic. Analyzing the network's nonlinear behavior allows engineers to anticipate and mitigate potential disruptions, improving its resilience and reliability. Similarly, in biological networks, understanding nonlinear dynamics is crucial for comprehending the functioning of the brain and for developing effective strategies for treating neurological disorders.

Nonlinear dynamics are not merely a complex concept; they are a fundamental aspect of network behavior with broad implications for various fields. From engineering and biology to social sciences, understanding nonlinearity is essential for comprehending the complexity of real-world networks. This understanding offers invaluable insights for designing efficient, robust systems, and for anticipating and mitigating potential disruptions, making it a critical area of study for anyone navigating the intricate world of network science. Ignoring this dynamism would be akin to trying to understand a river's course by only looking at a single drop of water; the true picture emerges only when considering the interconnectedness and the complex interactions of the whole. The journey into nonlinear dynamics might initially seem daunting, yet its rewards are immeasurable in terms of understanding and controlling complex systems. The tools and models are available to those willing to delve into this fascinating area of study.

2.1 Chaos and Complexity

Imagine a flock of birds, their movements seemingly chaotic, yet exhibiting a stunning, coordinated dance. This is a microcosm of chaos and complexity in networks, a realm where simple interactions between individual components give rise to intricate, unpredictable patterns. We've already covered the basics of networks – the nodes, edges, and simple structures. Now, we delve into the messy, beautiful unpredictability of nonlinear systems. Understanding this is crucial; these systems don't follow neat, predictable equations. Instead, they demonstrate sensitivity to initial conditions, the famous "butterfly effect." A tiny change in one part of the system can cascade into enormous, unforeseen consequences.
This isn't just theoretical mumbo-jumbo. Think about the stock market, a complex network of buyers and sellers. A single piece of news, a seemingly minor event, can send shockwaves through the entire system, leading to dramatic price swings. Or consider climate change, a complex network of interacting ecological systems. A slight increase in global temperature can trigger cascading effects – melting glaciers, rising sea levels, and altered weather patterns – far exceeding the initial perturbation. This sensitivity is a hallmark of chaotic systems, and it's vital to understand its implications.
Predicting the behavior of these systems is exceptionally challenging. Traditional linear models, those that assume a direct, proportional relationship between cause and effect, simply don't cut it. In

a nonlinear system, small inputs can yield disproportionately large outputs, and the system's response can be entirely unpredictable beyond a certain point. This is why simple forecasting models often fail in scenarios involving chaotic systems. One must use methods better suited for modeling this kind of complexity; this includes advanced numerical simulations, and advanced statistical techniques, focusing less on exact prediction and more on understanding the system's overall behavior and potential ranges of outcomes.

But chaos isn't just about unpredictability. It's also about the generation of rich, complex patterns. This emergent complexity is where the magic happens – where a multitude of simple interactions spontaneously give rise to sophisticated, organized behavior. The intricate branching patterns of a river system, the mesmerizing complexity of a snowflake, or the seemingly random distribution of stars in a galaxy – these are all manifestations of complex systems organizing themselves from basic rules and interactions. The sheer diversity of patterns that arise from this inherent complexity is remarkable. We find that even seemingly simple rules can generate complex and unpredictable behavior.

Yet, within this apparent randomness, there are often underlying structures. While a chaotic system might appear random at first glance, careful analysis can reveal hidden patterns and underlying order. These patterns often reveal themselves through the use of advanced mathematical and computational techniques like fractal geometry, which helps to quantify the self-similarity and scale-invariant properties of these complex systems. This highlights the dual nature of chaos: unpredictable in its details, yet possessing an underlying order in its overall structure and properties. It's a delicate balance.

The challenge lies in learning how to navigate and understand these complex systems. While precise prediction might be impossible, understanding the underlying dynamics and identifying key influencing factors can allow for more informed decision-making and risk management. The key is not to seek perfect prediction, but rather to develop a deep understanding of the system's behavior, its potential tipping points, and its range of possible outcomes. It's about shifting from a deterministic viewpoint to a probabilistic one, accepting the inherent uncertainty and leveraging the insights that come from understanding the system's dynamic interactions. This calls for approaches that embrace uncertainty and incorporate feedback loops, which continuously adapt to new information as it becomes available.

Understanding chaos and complexity is not merely an intellectual exercise; it holds profound implications for numerous real-world applications. From predicting market fluctuations to managing ecosystems to designing more resilient infrastructure, the ability to grasp the nuances of these nonlinear systems is becoming increasingly crucial. Moreover, the study of chaos and complexity serves as a bridge between different fields of study, uniting disciplines like mathematics, physics, biology, and economics under the umbrella of nonlinear dynamics. This interdisciplinary perspective enables a more holistic and comprehensive understanding of complex phenomena.

The tools for understanding nonlinearity, from advanced numerical techniques to novel mathematical frameworks, are constantly evolving. Researchers are constantly refining and developing new approaches to study chaos and complexity, constantly striving to gain a better understanding of the underlying mechanisms that drive the behaviour of these systems. The goal is to move beyond descriptive approaches and toward mechanistic understandings – to uncover the cause-and-effect relationships that drive these complex dynamics. This will ultimately lead to a deeper understanding of the world and better informed decision-making processes.

Let's reiterate, this chapter is not about predicting the future with perfect accuracy within these nonlinear systems. The inherent unpredictability of chaos must be accepted. The goal here is to equip ourselves with the conceptual tools and frameworks necessary to navigate the intricate landscapes of complex systems. By understanding their sensitivities, their emergent properties, and their underlying order amidst the apparent randomness, we can make more informed decisions, assess risks more effectively, and ultimately become more attuned to the dynamic world around us. This is the essence of mastering chaos.

2.2 Bifurcations and Transitions

Imagine a river, its flow seemingly constant, a gentle current meandering through its banks. Then, a sudden storm. The rainfall increases exponentially, swelling the river, changing its course. This, in essence, is a bifurcation–a point of significant change in a system's behavior, a transition from one stable state to another. In nonlinear networks, these bifurcations aren't always dramatic, like a raging flood; sometimes, they're subtle shifts, almost imperceptible at first, yet profound in their long-term consequences. Understanding these subtle shifts is crucial to comprehending the dynamics of complex systems.

These transitions aren't random events; they're governed by underlying parameters. Consider a network of neurons: the firing rate of individual neurons, the strength of their connections – these are all parameters that can push the network across a bifurcation point. A small increase in the synaptic strength, for instance, might trigger a cascade of activity, leading to a qualitatively different pattern of neural firing. This transition could manifest as a shift from a resting state to a synchronized oscillation, a dramatic change with significant implications for information processing.

The study of these bifurcations often relies on tools from dynamical systems theory. Phase diagrams, bifurcation diagrams, these mathematical constructs help us visualize the different regimes of behavior and the transitions between them. These tools aren't just theoretical exercises; they offer practical insights. By identifying the parameters that influence these transitions, we can gain a predictive understanding of network behavior. This predictive power is vital in various applications, from designing robust power grids to understanding the spread of disease in social networks. Anticipating these shifts is critical to mitigating their negative consequences.

Furthermore, consider the concept of hysteresis. In many nonlinear networks, the transition from one state to another isn't reversible. The path taken to reach a particular state influences the system's future behavior. Imagine the river again: once the flood subsides, the riverbed might have changed, its course permanently altered. Similarly, in a neural network, a strong stimulus might induce a persistent change in synaptic connectivity, creating a new stable state that may not easily revert to its previous condition. This phenomenon has profound implications for learning and memory.

This irreversibility adds a layer of complexity to understanding bifurcations. We need to consider not just the parameters that cause a transition, but also the history of the system. This is particularly important in networks with feedback loops, where the output of the system influences its future inputs. The interplay between feedback mechanisms and bifurcation points can lead to unexpected and often chaotic behavior. This inherent unpredictability highlights the importance of robust system design, capable of withstanding unexpected transitions.

The identification of critical transition points is not simply about observation. It necessitates a deep understanding of the network's structure and the interactions between its components. Analyzing network topology, studying the distribution of connection strengths, and identifying key nodes–these steps are essential for predicting and managing bifurcations. We can leverage computational models and simulations to probe the system's sensitivity to parameter changes. By systematically varying these parameters and observing the resultant transitions, we can build a comprehensive map of the network's dynamical landscape.

Beyond the purely theoretical, the practical implications of understanding bifurcations are extensive. Consider the design of resilient infrastructure networks–power grids, communication networks. Knowing the potential bifurcation points allows us to design systems that can withstand unexpected shocks and maintain

functionality even under stressful conditions. Similarly, in biological systems, predicting transitions is crucial for understanding disease progression and developing effective interventions. Early warning signals of impending transitions can be invaluable in these contexts.
The application of this knowledge transcends traditional boundaries. The principles of nonlinear dynamics and bifurcations apply equally to technological networks, biological systems, and social structures. This unifying framework allows us to address problems across various disciplines, creating a cross-pollination of ideas and methodologies. The beauty of network science lies precisely in its ability to connect seemingly disparate fields, offering a unified language to describe complex systems. The study of bifurcations represents a central pillar of this unifying framework, bridging the gap between theory and application. The ability to predict and manage these critical transitions is paramount to understanding and controlling the behavior of complex networks. This chapter, therefore, concludes with an emphasis on the ongoing importance of research in this field. The future of understanding complex systems hinges on our ability to accurately predict and effectively manage these fascinating and potentially disruptive transitions.

2.3 Synchronization Phenomena

Synchronization, the coordinated rhythmic behavior of coupled oscillators, is a ubiquitous phenomenon in nature and technology. Imagine a flock of birds, their movements seemingly choreographed in a breathtaking display. Or consider the rhythmic flashing of fireflies, a mesmerizing spectacle of natural synchrony. These are just two examples of how seemingly independent entities can spontaneously synchronize, forming a coherent collective pattern. This coordinated behavior isn't limited to biological systems; it also appears in technological networks, from power grids to coupled lasers, demonstrating the broad applicability of synchronization principles across diverse domains.

Understanding synchronization requires investigating the interplay between individual oscillators and their coupling mechanisms. Each oscillator has its intrinsic frequency, determining its natural rhythm. However, when oscillators interact through a network, their individual rhythms can be profoundly affected. The strength and nature of the coupling between them play a crucial role. Weak coupling may result in only partial synchronization, where some oscillators lock while others maintain their independent frequencies. Stronger coupling, on the other hand, can lead to complete synchronization, where all oscillators oscillate at the same frequency and phase.

The network topology itself, the pattern of connections between oscillators, significantly impacts the synchronization process. For instance, a highly connected network might synchronize more readily than a sparsely connected one. Certain network structures, such as those exhibiting small-world properties, have been shown to facilitate faster and more robust synchronization. This observation is particularly relevant to brain networks, where efficient information processing likely relies on the synergistic interplay of synchronized neural populations. Research in this area reveals intriguing connections between functional brain organization and the underlying anatomical connectivity patterns, emphasizing the importance of network topology in understanding brain function.

Several mathematical models provide frameworks for analyzing synchronization phenomena. The Kuramoto model, for instance, offers a simple yet powerful tool to study synchronization in large populations of coupled oscillators. This model, while simplified, captures the essence of synchronization dynamics and allows researchers to explore the influence of coupling strength, frequency distribution, and network topology on the emergence of synchrony. Variations and extensions of the Kuramoto model have been developed to address more complex scenarios, including situations where coupling is not uniform, or when oscillators have different natural frequencies. These advancements help researchers tackle real-world problems, such as optimizing the design of coupled oscillators for specific applications.

Beyond simple oscillators, synchronization manifests in more complex systems, such as coupled chaotic systems. The possibility of synchronizing chaotic systems seems paradoxical at first glance, as chaotic systems are inherently unpredictable. However, under specific coupling conditions, two or more chaotic systems can exhibit synchronized behavior, their unpredictable trajectories evolving in lockstep. This intriguing phenomenon has attracted significant attention, primarily due to its potential implications for secure communication and control systems. Using chaos synchronization, it's possible to transmit information securely, as the synchronized chaotic signal masks the transmitted data from eavesdroppers. The complexity of chaotic systems provides a natural barrier to unauthorized access.

The study of synchronization has far-reaching implications across multiple fields. In neuroscience, understanding synchronization helps us unravel the mysteries of brain function, including how different brain areas coordinate their activity to perform complex tasks. Moreover, disruptions in synchronization are implicated in various neurological disorders, highlighting the importance of synchronization for maintaining healthy brain function. In engineering, synchronization principles inform the design of robust and efficient systems, from power grids to communication networks. Maintaining synchronization in these systems is critical to their reliable operation and preventing catastrophic failures. Similarly, in ecology, synchronization plays a crucial role in the collective behavior of animal populations, influencing their survival and adaptation. Understanding these principles could offer valuable insights into conservation efforts and ecosystem management.

Finally, the ongoing research into synchronization highlights the intricate interplay between local interactions and global behavior. Individual oscillators, each behaving according to its own rules, can collectively generate surprisingly coordinated patterns. This emergence of order from chaos is a fundamental principle of complex systems, a testament to the power of collective dynamics. Further investigation into this fundamental behavior promises to unlock even deeper understanding across diverse fields and may hold the key to numerous yet-to-be-discovered applications. The potential for breakthroughs in this field is immense, encompassing scientific curiosity and technological advancement in equal measure. The future of synchronization research looks bright, with countless opportunities for new discoveries.

3 Neural Networks: Structure and Function

Biological neural networks, the intricate communication systems of our brains, serve as a compelling model for understanding information processing. Neurons, the fundamental units, communicate through electrochemical signals across synapses, forming a complex web of interconnected nodes. The strength of these connections, or synaptic weights, dynamically adjusts based on experience, a process known as synaptic plasticity, crucial for learning and memory. This adaptability allows the network to refine its responses and adapt to new stimuli, a feature absent in simpler systems. The remarkable computational power of the brain arises from the massively parallel processing enabled by this interconnected architecture. Understanding this biological blueprint provides a foundation for designing and improving artificial neural networks.

Artificial neural networks, inspired by their biological counterparts, provide a powerful tool for modeling complex systems and solving challenging computational problems. These networks consist of interconnected nodes, or artificial neurons, organized into layers: an input layer receiving data, hidden layers performing complex transformations, and an output layer providing the result. The connections between these neurons have associated weights, determining the influence of one neuron on another. Information flows through the network, undergoing transformations at each layer until a final output is produced. Unlike traditional algorithms, artificial neural networks excel in handling noisy or incomplete data, making them suitable for numerous applications. Moreover, their ability to learn from data is a defining feature.

The architecture of a neural network significantly influences its capabilities and applications. Feedforward networks, the simplest type, process information in one direction, from input to output. Recurrent networks, however, allow for feedback loops, enabling them to process sequential data and maintain internal states, making them suitable for tasks requiring memory, like language processing. Convolutional neural networks, with their specialized architecture, excel in image and video processing, efficiently extracting features from visual data. The choice of architecture hinges on the specific problem, and careful consideration is crucial for optimal performance. Innovative architectures continuously emerge, pushing the boundaries of neural network capabilities.

Learning in neural networks, the process by which they adapt to data, is central to their functionality. Supervised learning involves training the network on labeled data, where inputs are paired with desired outputs. The network adjusts its weights to minimize the difference between its predictions and the actual outputs, iteratively improving its accuracy. Unsupervised learning, on the other hand, deals with unlabeled data, allowing the network to discover underlying patterns and structures. Reinforcement learning involves training an agent to interact with an environment, learning through trial and error, optimizing its actions to maximize rewards. Each learning paradigm offers unique strengths, allowing for diverse applications.

The dynamics of neural networks, the intricate interplay of activation patterns and weight adjustments, are far from static. The network's behavior emerges from the interactions of its components, often exhibiting complex and unpredictable behavior. This dynamic nature is both a challenge and an opportunity. It allows for powerful adaptive behaviors, but also necessitates careful analysis to understand the network's function and ensure its stability.

The study of neural network dynamics involves analyzing the flow of information, identifying emergent patterns, and understanding how these patterns relate to the network's function. Research in this area seeks to develop new tools and techniques for analyzing and controlling these dynamic behaviors.

3.1 Biological Neural Networks

The human brain, a marvel of biological engineering, serves as the quintessential example of a complex network. Its intricate architecture, composed of billions of neurons interconnected by trillions of synapses, allows for the processing of information at an unparalleled scale. Each neuron, a tiny biological computation unit, receives input from numerous other neurons and, based on the summation of these inputs, fires an electrochemical signal. This signal, in turn, propagates along its axon to influence other neurons, forming a cascade of activity that underpins our thoughts, feelings, and actions. This intricate interplay of signals is the foundation of neural computation.

Understanding this biological system is key to unlocking the secrets of the brain and also forms a crucial cornerstone in the field of artificial neural networks. While artificial neural networks are designed to mimic some aspects of this biological complexity, they still differ significantly. The biological network is far more dynamic, adapting constantly through processes such as synaptic plasticity, which alters the strength of connections between neurons, influencing learning and memory. These adaptive mechanisms are far from fully understood, presenting a rich area for ongoing research and development. This dynamic adaptation is a crucial difference, and it's something that current artificial systems struggle to truly match.

Consider, for instance, the sheer diversity of neuronal types and their distinct roles within the brain's various regions. Not all neurons are created equal. Their varied morphologies and diverse signaling properties contribute significantly to the richness and complexity of neural computations. The integration of these different neuronal subtypes creates an emergent property that exceeds the simple sum of its parts. This is an area that further studies in network theory can shed more light on. It's not merely about the number of connections, but also about their specific types and the resulting functional heterogeneity. A deeper understanding of this diversity is crucial for advancing both theoretical neuroscience and practical applications.

Moreover, the brain's network structure isn't static; it is constantly undergoing rewiring and restructuring, a process often referred to as neuroplasticity. This continuous adaptation allows for learning, memory formation, and recovery from injury. The brain's response to stimuli and experiences is reflected in the changes in its underlying network structure and synaptic weights. These adaptive changes are not random but are precisely guided by specific patterns and rules; the underlying mechanisms driving these processes are still the subject of active investigation. However, it's this adaptive capability that makes biological neural networks so powerful and robust.

Furthermore, exploring biological neural networks offers unparalleled insights into the broader field of network science. Studying the brain's structure and function provides a unique lens for examining general network principles. For example, the brain's small-world architecture, characterized by high local clustering and short global path lengths, is a key organizational feature observed in various other complex networks, suggesting fundamental organizational principles that optimize information processing. The brain's efficiency in processing information could be a model for improving other types of networks. We can study its principles to understand how best to optimize a variety of different networks.

The complexity of biological neural networks extends beyond their structural features. The temporal dynamics of neural activity also play a vital role in computation. The precise timing of neuronal firing, and the intricate patterns of synchronized oscillations, influ-

ence information processing and contribute to cognitive functions. These synchronized patterns form dynamic modules that interact in a coordinated fashion. Understanding these dynamic interactions is essential for deciphering the brain's computational power. It's more than just a static wiring diagram; it's a symphony of activity, whose composition we are only beginning to grasp.
In conclusion, a comprehensive understanding of biological neural networks is pivotal for advancing both neuroscience and network science. The brain's intricate structure, dynamic adaptation, and sophisticated computational mechanisms provide a rich source of inspiration and guidance for designing more effective and robust artificial systems. By unraveling the intricacies of these biological networks, we gain invaluable insights that have implications far beyond the field of neuroscience alone, extending to areas such as artificial intelligence, robotics, and other complex system analysis. The ongoing study of these networks promises to unlock further breakthroughs in our understanding of complex systems, leading to innovations that impact various aspects of science and technology. The study of biological neural networks provides a unique opportunity to bridge the gap between biological and engineered systems, furthering the development of sophisticated and efficient technologies inspired by nature.

3.2 Artificial Neural Networks

Artificial neural networks (ANNs) mimic the structure and function of biological neural networks. They consist of interconnected nodes, or neurons, organized in layers. These layers process information through weighted connections, adjusting these weights during a process called training to improve performance on a specific task. The fundamental building block, the neuron, sums weighted inputs and applies an activation function to produce an output. This seemingly simple component, when replicated and interconnected, generates remarkable computational power.

Unlike traditional computing, ANNs excel at handling complex, nonlinear relationships. Imagine trying to program a computer to recognize a cat in a photograph. Defining explicit rules for all possible cat variations would be incredibly challenging, if not impossible. An ANN, however, can learn these complex patterns from a large dataset of cat images through a process of iterative adjustments to its connection weights. This adaptive nature distinguishes ANNs, making them particularly powerful in scenarios with ambiguous or incomplete information.

The architecture of an ANN significantly impacts its capabilities. Feedforward networks, the most basic type, propagate information in one direction, from input to output. Recurrent networks, on the other hand, have feedback loops, enabling them to retain information and process sequences, making them suitable for tasks involving time-dependent data, like natural language processing. Convolutional neural networks (CNNs) excel at processing grid-like data, such as images and videos, employing convolutional layers to extract spatial features efficiently. The choice of architecture depends heavily on the nature of the task and the characteristics of the input data.

Training an ANN involves adjusting the weights of its connections to minimize errors between predicted and actual outputs. This is usually achieved using backpropagation, an algorithm that calculates the error gradient and iteratively updates the weights to reduce it. Several training strategies exist, including stochastic gradient descent, which updates weights based on small batches of data, and Adam, an adaptive optimization algorithm that dynamically adjusts the learning rate. Proper selection of the training algorithm and its parameters is crucial for achieving optimal performance and avoiding issues like overfitting, where the network memorizes the training data rather than learning generalizable patterns.

Despite their power, ANNs are not without limitations. They often require vast amounts of data for training, and their internal workings can be opaque–sometimes referred to as the "black box"

problem—making it difficult to understand why they arrive at specific conclusions. This lack of transparency poses challenges in areas like medicine or finance, where understanding the reasoning behind a prediction is critical for trust and accountability. Addressing interpretability and improving efficiency with limited data remain active areas of research.
Furthermore, the computational cost of training large ANNs can be substantial, requiring specialized hardware such as GPUs to accelerate the process. The energy consumption associated with training these networks raises environmental concerns, prompting research into more energy-efficient architectures and training techniques. Ongoing research explores the development of more efficient architectures and algorithms to mitigate these limitations, pushing the boundaries of what ANNs can achieve.
Beyond the challenges, the potential applications of ANNs are vast. They are currently used in image recognition, natural language processing, speech recognition, medical diagnosis, and financial modeling, just to name a few. Their ability to learn complex patterns from data opens up new possibilities in scientific discovery and technological innovation. This adaptability across diverse domains exemplifies their value as powerful tools for solving intricate problems. ANNs are a crucial element in the broader field of network science, bridging the gap between biological and artificial systems. Their continued development promises further breakthroughs, pushing the limits of computation and impacting various aspects of life in unexpected ways. The future development of this technology will likely be driven by advancements in hardware and algorithmic efficiency, allowing the application of this computational power to previously intractable challenges.

3.3 Network Architectures

Understanding network architectures is crucial for building effective neural networks. We've already covered the fundamentals of neural networks, but now it's time to delve into the diverse range of structures that underpin their power. The choice of architecture significantly impacts a network's capabilities, determining its ability to learn specific tasks and patterns. A poorly chosen architecture can lead to suboptimal performance, even with the best training data and algorithms.

Consider a simple feedforward network, where information flows unidirectionally from input to output layers, passing through hidden layers. This architecture is straightforward, ideal for tasks involving straightforward mappings between inputs and outputs. Its simplicity allows for efficient training, making it a popular choice for beginners. However, its limitations become apparent when dealing with complex, sequential data, or tasks requiring memory of past inputs.

Recurrent neural networks (RNNs) offer a solution to this limitation. Unlike feedforward networks, RNNs possess loops, allowing information to persist and influence future computations. This internal memory makes them exceptionally well-suited for processing sequential data, such as time series or natural language. Different RNN architectures exist, each with its strengths and weaknesses. Long Short-Term Memory (LSTM) networks, for instance, address the vanishing gradient problem often encountered in standard RNNs, allowing them to learn long-range dependencies within sequences. Gated Recurrent Units (GRUs) offer a simpler alternative, providing a good balance between performance and computational efficiency.

Convolutional Neural Networks (CNNs) represent a significant departure from RNNs and feedforward architectures. Their strength lies in their ability to extract features from spatial data, such as images or videos. CNNs employ convolutional layers that apply filters to the input, effectively detecting patterns at different scales and locations. This hierarchical feature extraction makes them exceptionally powerful for tasks like image recognition, object detection, and image segmentation. The depth of a CNN, the number of convolutional layers and filters, greatly influences its ability to

learn complex visual features.
Graph Neural Networks (GNNs) are designed specifically to handle data structured as graphs. These networks operate on nodes and edges, learning representations that capture relationships between connected entities. This makes them ideal for tasks involving social networks, molecular interactions, or any data that can be represented as a graph. Different GNN architectures, such as Graph Convolutional Networks (GCNs) and Graph Attention Networks (GATs), employ different methods to aggregate information from neighboring nodes, resulting in varying performance characteristics.
The selection of an appropriate architecture is not a trivial task. It necessitates a comprehensive understanding of the problem at hand, the nature of the data, and the computational resources available. Overly complex architectures might lead to overfitting, while simplistic designs might be incapable of capturing the intricacies of the data. Therefore, careful consideration and experimentation are essential in choosing the best architecture for a given task. It's often necessary to try several architectures before finding the optimal one, as there is no "one-size-fits-all" solution.
Furthermore, hybrid architectures, combining elements from different network types, are becoming increasingly prevalent. These combinations leverage the strengths of individual architectures, creating highly powerful and versatile neural networks capable of addressing complex and multifaceted problems. For instance, a hybrid model could combine convolutional layers for feature extraction from images with recurrent layers for processing sequential information within the images, like identifying the actions of objects within a video. The possibilities are vast and constantly evolving, reflecting the vibrant and dynamic nature of this field.
Finally, remember that the architecture is only one piece of the puzzle. The choice of training algorithm, hyperparameters, and data preprocessing techniques all play crucial roles in achieving optimal performance. Even with the best architecture, suboptimal training can lead to poor results. A comprehensive understanding of all

these factors is crucial for building effective and high-performing neural networks. Through careful consideration and experimentation, you can unlock the full potential of these powerful tools. The journey from understanding the basics to mastering these architectures requires dedication and continuous learning. Embrace the challenges, experiment boldly, and let your curiosity guide you.

3.4 Learning and Adaptation

The ability of neural networks to learn and adapt is central to their power. This process, however, isn't a simple on/off switch; it's a nuanced dance between adjusting internal parameters and responding to external stimuli. Think of it like a sculptor refining a masterpiece: each adjustment is deliberate, guided by the desired outcome, but the overall form emerges organically. We'll explore the mechanisms behind this learning process, focusing on both the biological inspiration and the engineered implementations.

Understanding biological neural networks offers crucial insight. Neurons, the fundamental building blocks, communicate via electrochemical signals across synapses. The strength of these connections, the synaptic weights, are dynamically adjusted based on the activity patterns. This process, known as synaptic plasticity, is the cornerstone of learning in the brain. Hebb's postulate, "neurons that fire together wire together," elegantly captures this core principle: repeated co-activation strengthens the connection between neurons, creating pathways for efficient information processing. Long-term potentiation (LTP) and long-term depression (LTD) represent specific mechanisms underpinning this plasticity, finely tuning the network's response to repeated stimuli. These biological processes aren't just interesting; they're the blueprints for many artificial learning algorithms.

Artificial neural networks mimic this biological elegance. They employ various learning rules to adjust their parameters, seeking to minimize error and improve prediction accuracy. Supervised learning, a common approach, uses labelled datasets to guide the network's parameter adjustment. The backpropagation algorithm, for example, propagates errors backward through the network, adjusting the weights to reduce the difference between the network's output and the target value. This iterative process refines the network's ability to generalize, to accurately predict outcomes for unseen data. However, achieving this generalization requires careful consideration of factors like network architecture, training data selection, and the prevention of overfitting, a common pitfall in which the network learns the training data too well, failing to generalize to new instances.

Unsupervised learning, conversely, operates without labelled data. Here, the network seeks patterns and structures within the data itself. Self-organizing maps and clustering algorithms are examples. These methods are powerful for data exploration and dimensionality reduction, allowing us to uncover hidden structures and relationships in high-dimensional datasets. Reinforcement learning, another significant paradigm, trains agents to interact with an environment and learn optimal strategies through trial and error. The agent receives rewards for desired actions, learning to maximize its cumulative reward over time. This methodology is especially valuable for complex control problems and game playing, where the optimal strategy isn't readily apparent.

Adaptability is paramount. The world is dynamic, and effective networks must adjust to changing conditions. Incremental learning allows networks to incorporate new information without retraining on the entire dataset, offering efficiency and scalability. Catastrophic forgetting, however, poses a challenge: networks can sometimes overwrite previously learned information when exposed to new data. This problem necessitates strategies that preserve previously acquired knowledge while accommodating new information. Regularization techniques, such as weight decay or dropout, help to mitigate overfitting and improve generalization performance, enhancing the network's long-term adaptation capabilities.

Therefore, the learning and adaptation capabilities of neural networks are not static attributes. They are outcomes of carefully designed architectures and learning algorithms, inspired by biological processes and refined through rigorous engineering. The choice of learning paradigm, from supervised to unsupervised to reinforcement, fundamentally shapes the network's behavior and its ability to process and respond to information. Furthermore, the ongoing development of novel learning algorithms and architectural refinements continues to expand the capabilities and applicability of these remarkable systems. Mastering the intricacies of learning and adaptation is essential to unlocking the full potential of neural networks across countless applications.

3.5 Neural Network Dynamics

Understanding neural network dynamics is crucial for comprehending their behavior and capabilities. These networks, whether biological or artificial, are fundamentally dynamic systems, constantly evolving and adapting in response to input and internal processes. Their behavior is not simply a static mapping of inputs to outputs; instead, it's a complex interplay of interconnected nodes, each influencing and being influenced by its neighbors. This constant interaction leads to a rich spectrum of emergent properties, not easily predictable from the behavior of individual components.

The temporal evolution of neural network activity is a key aspect of their dynamics. Consider a simple feedforward network. While the processing may appear instantaneous at a macroscopic level, microscopic examination reveals a cascade of activations spreading across the network, a wave of excitation traveling through interconnected layers. Each neuron's firing pattern, its timing, and the precise sequence of activation contribute significantly to the network's overall output. Understanding these temporal dynamics requires analyzing the interplay between excitatory and inhibitory connections, considering factors such as synaptic delays and the

inherent time constants of neuronal membranes. This detailed, temporal analysis moves beyond simple snapshots of network state, revealing the critical role of timing in information processing.
Recurrent neural networks display even more intricate dynamics. These networks, with their feedback loops, exhibit dynamic behavior far exceeding the simple feedforward case. Their state evolves over time, influenced not just by current inputs, but also by their past activations. This "memory" allows recurrent networks to process sequential data, model temporal dependencies, and generate complex patterns. Think of language models: their internal states reflect the context of previous words, enabling them to predict the next word in a sentence. This temporal context shapes the very essence of their processing, highlighting the profound importance of understanding recurrent network dynamics.
Moreover, the stability and robustness of neural networks, crucial for their reliable operation, are intrinsically linked to their dynamics. A stable network maintains consistent activity patterns in response to similar inputs; a robust one continues functioning even when facing perturbations. These properties depend on the network's architecture, the strength and distribution of connections, and the nature of the learning algorithms used to train it. For instance, chaotic behavior, often seen in highly connected networks, can be detrimental, making prediction difficult and rendering the system unreliable. In contrast, controlled oscillations, a form of dynamic synchronization, can be crucial for coordinated activity and efficient information processing. Thus, controlling and shaping the dynamics of neural networks is crucial for engineering reliable and effective systems.
Investigating the dynamics of neural networks often involves employing mathematical tools. Differential equations can describe the evolution of neuron activations over time, capturing the continuous nature of neuronal responses. However, the complexity of these systems often necessitates the use of computational simulation, allowing for the exploration of large-scale networks and the analysis of emergent behavior that defies analytical solutions. Such

simulations provide valuable insights into the impact of network parameters on the overall dynamics, illuminating the complex relationship between structure, function, and emergent behavior in these dynamic systems. Combining mathematical modeling with computational simulations allows researchers to explore and predict the behavior of neural networks in different scenarios.

In conclusion, a deep understanding of neural network dynamics is vital for realizing their full potential in various applications. From analyzing the subtle interplay of activations in feedforward architectures to modeling the complex temporal dependencies inherent in recurrent networks, unraveling the intricate dynamics is crucial. The stability, robustness, and overall performance of these networks hinge on these dynamics. Whether through mathematical modeling, computational simulations, or a combination of both, a continuous effort to understand these dynamic processes will unlock even more efficient and powerful neural network architectures in the years to come. The exploration of these complex systems reveals not just intricate patterns of computation, but also fundamental insights into the mechanisms underlying information processing in both biological and artificial systems.

4 Connectivity and Information Flow

Understanding how information flows through a network is crucial; it dictates the network's efficiency, its resilience, and its overall function. Think of a bustling city: its roads represent connections, the cars are information packets, and traffic jams are bottlenecks in information flow. The city's design–its network topology–directly influences how smoothly (or not) things move. A grid-like structure might be efficient for some types of flow, but vulnerable to disruptions. A more decentralized, random network might be more robust, but less efficient for targeted delivery.

Network topology plays a significant role, shaping how information propagates. A scale-free network, with a few highly connected hubs and many sparsely connected nodes, facilitates rapid information spread, as information can quickly reach these hubs. However, this also makes it vulnerable: damaging a hub can severely disrupt the entire network. Conversely, a regular lattice network, with each node having the same number of connections, offers more resilience but potentially slower information dissemination. Consider the impact of this on, say, disease transmission or social influence.

Information propagation models provide a framework for analyzing this process. The simplest model, perhaps, assumes that information spreads directly from connected nodes. More complex models account for factors like node properties—perhaps some nodes are more influential than others—and the probability of successful information transmission. The likelihood of information reaching a given node depends intricately on the network's structure and the characteristics of both the information itself and the nodes it encounters. Some information might decay rapidly, while other, more robust ideas, persist and spread wider.

The speed of information flow is influenced by factors beyond mere topology. The nature of the information itself plays a significant part. Complex information might require more processing time at each node, slowing the overall process. Similarly, the capacity of each connection (the bandwidth, if you will) is critical. A network with high-capacity links can handle a greater volume of information simultaneously. Think of a high-speed internet connection versus a dial-up modem—the difference in speed is striking.

Beyond simply the speed and reach, we must consider the quality of information flow. Distortion and misinformation can arise as information travels across the network. This is especially true in large, complex networks where multiple paths exist between nodes. Errors can accumulate along the way, leading to a degraded or even completely misleading final product. Mechanisms for error correction and verification become essential as the network grows in size and complexity.

Furthermore, the dynamic nature of many networks—connections forming and breaking, nodes joining and leaving—adds another layer of complexity. Real-world networks aren't static entities; they constantly evolve. This dynamism affects the flow of information in profound ways. A sudden disruption—the loss of a key node or a connection failure—can significantly alter the network's efficiency. It's important to understand these fluctuations and their impacts on network behavior.

To better understand and manage information flow, tools and techniques from network science offer invaluable insights. Network analysis, simulation, and modeling allow us to study the patterns and dynamics of information propagation in various types of net-

works. We can test different network designs, predict the spread of information, and identify vulnerabilities. By understanding information flow mechanisms, we can improve the efficiency and robustness of various real-world systems.
In conclusion, analyzing the connectivity and information flow within a network requires understanding the interplay between network topology, information characteristics, and network dynamics. By examining these factors through the lens of network science, we gain the ability to model, predict, and ultimately optimize the transfer of information within these intricate systems. This knowledge allows us to design more efficient and resilient networks across countless applications, from technological infrastructures to biological systems. The implications are vast, impacting fields as diverse as communication, epidemiology, and societal influence. It's a deep dive that reveals fundamental principles governing the complex interplay of connection and information.

4.1 Network Topology

Network topology dictates a network's overall structure, profoundly impacting its functionality and behavior. Understanding this structure is crucial for analyzing information flow, predicting network resilience, and designing efficient systems. Consider, for instance, a completely connected network–every node linked to every other. While seemingly robust, this topology becomes computationally expensive and inefficient as the network grows, rendering it impractical for large-scale systems.
Conversely, a star topology, with a central hub connecting to all other nodes, presents a different set of characteristics. Information transmission is efficient, as it typically passes through the central node. However, this central node becomes a single point of failure; if it malfunctions, the entire network collapses, highlighting a critical vulnerability. Therefore, the choice of topology is not merely a design consideration but a fundamental determinant of

the network's strengths and weaknesses.

Think about the human brain. Its intricate network of neurons is far from a regular lattice or a simple star; its topology is complex, exhibiting both local clustering and long-range connections, a hallmark of small-world networks. This specific architecture enables both efficient information processing within localized regions and rapid communication across disparate brain areas, crucial for cognitive functions. The brain's topology is not static; it dynamically reconfigures itself through learning and experience, adapting its connectivity to meet changing demands. This adaptability is a key aspect of its remarkable resilience and flexibility.

Several key topological features help us understand and characterize networks. Degree distribution, indicating the number of connections per node, provides insight into the network's overall structure. A scale-free network, for example, exhibits a power-law degree distribution, characterized by a few highly connected "hub" nodes and many sparsely connected nodes. This peculiar architecture contributes to the network's resilience, as the removal of many sparsely connected nodes has a minimal impact on its functionality. However, the removal of a hub node can be catastrophic.

Path length, the shortest distance between two nodes, determines the efficiency of information transmission. In small-world networks, characterized by short average path lengths despite high clustering, information can travel rapidly across the network. This efficient communication is a critical feature of many real-world networks, including social networks and the Internet, enabling swift information dissemination and rapid response to changes. Understanding these path lengths, therefore, allows prediction of information spread and potential bottlenecks within a network.

Clustering coefficient, measuring the interconnectedness of a node's neighbors, reflects the network's local structure and information flow patterns. High clustering suggests strong local interactions, influencing the spread of information and the network's stability. Analyzing clustering patterns unveils the network's community structure – groups of densely interconnected nodes—critical for understanding social dynamics, information cascades, and the resilience of certain network sub-structures.

Beyond these fundamental topological properties, researchers also examine network motifs, recurring patterns of interconnections, to identify functional units and predict network behavior. These motifs, small subgraphs that appear significantly more often than expected by chance, offer valuable insights into the design principles governing the network's organization and can even point towards specific functional roles. Identifying such motifs can help researchers understand not only the overall structure but also the specific roles and interactions of individual components within complex systems.

Different types of networks exhibit distinct topological properties, reflecting their underlying mechanisms and functions. While random networks offer a baseline for comparison, exhibiting a Poisson degree distribution and short path lengths, real-world networks often deviate considerably from this idealized model. This deviation highlights the importance of considering specific topological features when analyzing real-world systems, whether it's a biological neural network, a social network, or a technological infrastructure. The choice of appropriate metrics and analytical tools must reflect this complexity.

In conclusion, network topology is not a static property but a dynamic interplay of interconnected nodes and edges, shaping information flow and influencing the network's resilience and efficiency. A deep understanding of these topological features is essential for both theoretical advancements and practical applications. From analyzing the robustness of critical infrastructures to designing efficient communication systems or understanding the dynamics of biological systems, a grasp of network topology is crucial for navigating the complexities of interconnectedness in the modern world. The tools and approaches discussed in this chapter provide a foundation for exploring these complexities and extracting valuable insights from diverse network systems. Successfully navigating this landscape requires a blend of theoretical understanding and practical application of these concepts, allowing for informed decisions regarding network design, analysis, and control.

4.2 Information Propagation

Understanding how information spreads through a network is crucial. It's not just about the speed; it's about the pathways, the bottlenecks, and the overall efficiency. Think of it like a river system: a wide, slow-moving river might carry a lot of water, but a smaller, faster-flowing stream might deliver information to a specific location much quicker. The network's topology–its structure–dictates these pathways, determining how quickly and broadly information disseminates.

Consider a social network. A highly centralized network, like a star, where one person is connected to many others, allows that central node to distribute information swiftly. However, if that central node becomes unavailable–say, their account is suspended–information flow grinds to a halt. Conversely, a decentralized network, like a mesh, is robust; losing a few nodes has less impact on overall connectivity and information propagation. This redundancy built into the system is key. The resilience of a network against disruptions is directly related to how information moves through it.

The type of information also matters. A short, easily digestible message might spread rapidly through various channels, like viral content on social media. In contrast, complex information requiring careful consideration and analysis might travel more slowly, moving along pre-established pathways within more tightly knit communities. The nature of the information itself–its inherent attractiveness or importance–influences its propagation. Consider how rumors or misinformation can spread like wildfire compared to factual, evidence-based information. The latter often requires more effort to disseminate.

Modeling information propagation requires sophisticated tools.

Epidemiological models, originally designed to track disease outbreaks, are remarkably effective in illustrating information spread. These models capture the dynamics of infection, which can be likened to the spread of ideas or messages through a network. They use parameters like infectivity (how easily a message is passed on) and recovery (how quickly someone stops sharing it) to predict the spread's trajectory. They allow us to investigate different scenarios and how changing network properties or communication patterns affects the outcome.

Furthermore, the concept of network centrality takes on new meaning in this context. A highly central node is not only important for its connections but also for its influence on information flow. Nodes with high betweenness centrality, for example, control the flow of information between different groups within the network. They essentially act as bridges. Understanding these central nodes allows for targeted strategies to either facilitate or impede information spread, whether for public health campaigns or managing misinformation. Therefore, identifying key individuals and understanding their position in the network becomes crucial for influencing communication patterns.

Real-world applications are abundant. Think about the spread of news, the diffusion of innovations, or the propagation of marketing messages. Understanding these processes isn't just an academic exercise; it's essential for effective communication strategies across many sectors. Businesses use this to target their advertising; government agencies employ it for public health initiatives and disaster response; and social scientists use it to analyze trends and societal changes. Therefore, recognizing the influence of network structure and information characteristics is essential for successfully navigating the complex world of information propagation.

Analyzing information flow involves more than just observing the spread itself; we need to consider the interplay between content and structure. The effectiveness of the information dissemination relies greatly on its coherence and resonance with the audience. The most impactful messages adapt to the network's structure, leveraging existing pathways and connections to reach their target audience. Therefore, efficient propagation isn't just about speed but is also about effective targeting and the right message for the right audience. It's a delicate balancing act.

The ability to anticipate and possibly control information flow is not merely academic curiosity, it is also a significant endeavor. Understanding the mechanisms of information propagation can allow for the development of intervention strategies. If a false narrative is spreading rapidly, this knowledge allows targeted countermeasures to correct it. Similarly, if positive news is failing to reach its intended audience, strategic adjustments to the dissemination approach can prove beneficial. This capability to understand and shape information flow has enormous societal impact and offers tremendous implications for different sectors.

To conclude, understanding information propagation within complex networks isn't simply a matter of tracing paths. Instead, it requires a deep understanding of network topology, information characteristics, and the intricate interplay between them. It involves using sophisticated models to simulate information spread, identifying key influencers, and designing targeted intervention strategies. This knowledge proves vital in various sectors, shaping everything from marketing campaigns to public health interventions. The power to understand and potentially control this flow has significant implications for societies across the globe.

5 Emergent Phenomena in Complex Networks

Consider a flock of birds, each reacting to its immediate neighbors. Individually, their movements appear erratic, almost random. Yet, collectively, they form breathtakingly coordinated patterns, swirling and shifting as one. This seemingly simple interaction, at the local level, gives rise to a complex, emergent behavior at the global scale. This is the essence of emergent phenomena in complex networks. It's not planned, it's not centrally controlled, it simply emerges from the intricate dance of many individual interactions.

This chapter focuses on these unpredictable yet often predictable patterns, these unexpected but not entirely surprising outcomes. We'll explore how these collective behaviors arise from the fundamental interactions within networks, delving into the mechanics of self-organization and the robustness that complex systems often display. Understanding these emergent properties is crucial, not just for theoretical advancements, but also for practical applications in various fields.

Let's delve into self-organization. Imagine a colony of ants building a complex nest without a central architect. Each ant follows simple rules: detecting pheromones, carrying debris, responding to stimuli. These local rules, when combined across the entire colony, result in the creation of a highly structured and functional nest. This isn't a conscious decision; it's an emergent property arising from the decentralized interactions of numerous individual agents. This concept extends far beyond ant colonies. Think of the intricate patterns in biological systems, the formation of crystals, or even the spontaneous synchronization of fireflies flashing in unison. These are all examples of self-organization in action, demonstrating how complex structures can arise from relatively simple local interactions.

Next, let's consider network resilience. Real-world networks–from the internet to the power grid to the human brain–are constantly subjected to disturbances: equipment failures, attacks, natural disasters. The ability of these networks to withstand such disruptions and maintain their functionality, is directly linked to their inherent structure and the nature of the interactions between their components. A highly centralized network, for instance, might be vulnerable to a single point of failure. Conversely, a decentralized network, with its numerous redundant pathways, might be considerably more resilient. This robustness is not inherent to the individual components, but an emergent property of the network's overall architecture. Research has shown the importance of network topology–the arrangement of connections–in determining resilience. Understanding this relationship is critical in designing robust and reliable systems. A network's resilience is not solely dependent on its topology; it is also influenced by the dynamics of information flow and the adaptability of individual nodes. Imagine a network where nodes can dynamically reroute information or adapt their functions based on current conditions. Such adaptive networks exhibit a heightened resilience compared to static networks. This adaptive capacity is itself an emergent property of the network.

The study of emergent phenomena is therefore not just about identifying patterns; it's about understanding the underlying mechanisms that give rise to them. It's about exploring how local interactions translate into global behavior, and how seemingly simple rules can lead to complex, unpredictable outcomes. By studying these phenomena, we gain invaluable insights into the nature of complexity itself, leading to a deeper understanding of various systems ranging from biological networks to social structures and technological infrastructures. This understanding allows for the prediction of system behavior and informs the design of more robust and adaptable systems in diverse fields. The implications are far-reaching, impacting everything from disaster preparedness and public health to the development of more resilient infrastructure and efficient algorithms. Furthermore, emergent phenomena are not confined to static systems; they also manifest in dynamic environments, adding another layer of complexity to their study. The continuous evolution of network structures and the interactions between nodes further contribute to the emergence of intricate patterns and behaviors. The ever-changing nature of these systems demands advanced analytical techniques capable of capturing and interpreting these transient phenomena.

Therefore, the study of emergent phenomena in complex networks presents a rich and challenging field of research with substantial practical implications. The ability to predict and manipulate these phenomena holds the key to designing more robust, efficient, and adaptive systems across a wide range of applications. Understanding how these emergent properties arise is not just intellectually stimulating; it's fundamentally important for building a more resilient future. From optimizing infrastructure to safeguarding against epidemics, the insights gleaned from studying emergent behavior hold immense value. Continual research in this area is essential for navigating the complexities of our increasingly interconnected world and harnessing the power of emergent phenomena for the benefit of society.

5.1 Collective Behavior

Imagine a flock of birds, seemingly chaotic yet moving in perfect unison. This elegant, coordinated movement isn't directed by a central leader; instead, it arises from simple rules followed by each individual bird, reacting to its immediate neighbors. This is a prime example of collective behavior, a phenomenon where the interactions of individual units generate complex, emergent patterns at the system level. These patterns often exhibit properties that are far from obvious from looking at the individual components alone. The study of collective behavior necessitates understanding the delicate balance between local interactions and global outcomes. This is especially relevant in complex networks, where the interplay between individual nodes and their connections leads to surprising system-wide characteristics.

Consider a social network, for instance. Each individual's actions–their likes, shares, and comments–represent local interactions. However, these seemingly small actions can lead to large-scale trends, viral memes, or even social movements. Understanding how these local interactions cascade and propagate throughout the network is crucial for comprehending the overall dynamics of the social system. Predicting such emergent phenomena requires advanced modeling techniques that account for the intricate network structure and the nonlinear nature of social interactions. Failure to consider these factors often leads to inaccurate predictions, highlighting the need for sophisticated computational tools and analytical frameworks.

The emergence of collective behavior in complex networks is not simply an interesting theoretical concept; it has substantial practical implications. In fields like epidemiology, understanding how diseases spread through networks is paramount for developing effective containment strategies. Similarly, in transportation systems, analyzing traffic flow as a collective behavior allows for the optimization of traffic management and the prediction of potential bottlenecks. Even in financial markets, the collective behavior of investors can lead to market crashes or booms, emphasizing the need to understand these dynamics to mitigate risks. This underscores the critical role of collective behavior analysis in various domains, ranging from public health to economic policy.

However, analyzing collective behavior isn't straightforward. The inherent complexity of many networks necessitates the use of sophisticated mathematical tools and computational methods. Agent-based modeling, for example, allows researchers to simulate the behavior of individual agents within a network and observe the emergent patterns that arise from their interactions. This approach has proven particularly useful in studying systems where analytical solutions are intractable, capturing the richness and nuance of complex interactions. The careful consideration of network topology, the specific rules governing individual interactions, and the inherent noise in the system is crucial for accurate and insightful modeling.

Moreover, studying collective behavior requires a multidisciplinary approach. Researchers draw on principles from physics, mathematics, computer science, and the social sciences to develop comprehensive models and analytical tools. This interdisciplinary nature underscores the complexity of the phenomenon and the need for collaborations between experts from diverse fields. The development of new theoretical frameworks and the refinement of existing ones are ongoing processes, constantly evolving with technological advances and new research findings. This ongoing effort emphasizes the vibrant and dynamic nature of the field.

The complexity of collective behavior makes precise prediction challenging. While models can provide valuable insights into the overall dynamics, they often fall short of predicting specific outcomes. This inherent unpredictability underscores the importance of focusing on understanding the underlying mechanisms and general trends rather than seeking absolute accuracy in predictions. Instead of striving for perfect forecasting, a more practical approach involves identifying critical factors that influence collective behavior and assessing the relative likelihood of different outcomes. This probabilistic perspective offers a more nuanced and realistic approach to studying complex systems. This perspective necessitates a shift away from deterministic models towards models that explicitly incorporate randomness and uncertainty.

In conclusion, the study of collective behavior in complex networks is a rapidly evolving field with significant implications across various domains. Understanding how local interactions lead to global patterns is crucial for tackling challenges ranging from predicting disease outbreaks to managing traffic flow. While the inherent complexity of these systems presents substantial challenges, the use of sophisticated modeling techniques and a multidisciplinary approach holds the key to unlocking valuable insights and developing effective strategies for managing and harnessing the power of collective behavior. Continued research and interdisciplinary collaboration remain essential for advancing our understanding of this fascinating and intricate phenomenon. Further investigations are required to better understand the interplay between individual actions, network structures, and emergent global behaviors, paving the way for innovative solutions in diverse fields. The potential applications are vast and the challenges stimulating, making this area of study both intellectually rewarding and practically relevant.

5.2 Self-Organization

Self-organization is a fascinating phenomenon where intricate structures and patterns emerge from simple interactions within a network. Imagine a flock of birds, seemingly chaotic yet exhibiting coordinated movements. This is a prime example of self-organization, a process driven by local rules leading to global order. No central controller dictates their actions; instead, each bird responds to its immediate neighbors, creating the mesmerizing collective behavior we observe. This principle, where local interactions give rise to global order, is crucial in understanding the complexity of numerous systems, from ant colonies to the human brain. Understanding these principles allows us to grasp the emergent properties of complex networks, a key theme of this book.

The concept of self-organization is deeply rooted in the principles of nonlinear dynamics, a topic explored earlier. Remember the discussion on bifurcations and transitions? In self-organizing systems, we often observe these critical points where small changes in initial conditions can dramatically alter the resulting patterns. This sensitivity underscores the unpredictable yet fascinating nature of self-organization. A slight variation in the interactions between individual components can lead to vastly different overall structures, highlighting the inherent unpredictability and nonlinearity of these systems. The beauty lies in the emergent order arising from this seeming chaos.

Consider the development of a complex network like the internet. No central authority planned its intricate structure. Instead, it emerged through decentralized decisions by countless individuals and organizations connecting and reconnecting. This spontaneous formation, driven by individual needs and local interactions, exemplifies the power of self-organization in shaping complex systems. The robustness and adaptability of the internet partly stem from this inherent self-organizing nature, allowing it to adapt to disruptions and continue functioning even in the face of challenges.

The implications of self-organization extend beyond technological networks. Biological systems, such as the intricate structure of the human brain, also exhibit remarkable self-organizing properties. Neurons, the fundamental building blocks of the brain, communicate through complex interactions, leading to the emergence of higher-level cognitive functions. The development and plasticity of the brain, its ability to adapt and learn, are intimately linked to the self-organizing processes within its neural networks. This self-organization, in turn, shapes our thoughts, memories, and actions, emphasizing the profound impact of this process on living organisms.

Moreover, the study of self-organization offers valuable insights into the resilience and robustness of complex networks. Understanding how self-organizing systems adapt to changes and disturbances is crucial for designing robust and resilient systems in various domains, from engineering to ecology. By analyzing the underlying mechanisms of self-organization, we can develop more resilient systems that can withstand unexpected shocks and continue to function effectively. These insights are invaluable in designing critical infrastructures, such as power grids and communication networks, which must maintain functionality in the face of unforeseen circumstances. The key lies in recognizing the principles that allow these systems to self-organize and adapt.

In essence, self-organization is a powerful process driving the emergence of complex structures and patterns in a myriad of systems. It's a fundamental principle shaping the evolution of everything from ant colonies and the internet to the intricate networks within the human brain. By understanding its mechanics, we can unlock valuable insights into the behavior, robustness, and adaptability of complex systems, allowing us to design more resilient and efficient systems for the future. This understanding, rooted in the principles of nonlinear dynamics and network science, paves the way for innovative applications across diverse fields, from engineering and computer science to biology and social sciences. The exploration of self-organization remains a crucial area of research, constantly revealing new facets of complexity and offering insights into the intricate interconnectedness of the world around us. The implications of this understanding are far-reaching, influencing how we design, build, and manage systems in a world increasingly characterized by complex, interconnected networks. The study of self-organization is not merely an academic pursuit; it is a powerful tool for understanding and shaping our future.

5.3 Resilience and Robustness

Understanding a network's resilience means comprehending its ability to withstand disruptions and maintain functionality. This isn't merely about surviving a single shock; it's about adapting and recovering from various stresses, whether they're sudden failures, gradual degradation, or even intentional attacks. Imagine a power grid facing a sudden surge – a resilient grid will reroute power, minimizing outages. This capacity hinges on several factors, which we will explore in detail.

Robustness, a closely related concept, goes beyond mere survival. A robust network not only withstands shocks but also maintains its core functions and performance even under stress. Think of the internet; while individual servers might fail, the overall functionality remains largely intact thanks to redundant pathways and robust protocols. The key lies in understanding and designing for redundancy and distributed control. Designing for robustness involves anticipating potential failure points and developing systems to mitigate their impact.

Network topology plays a crucial role in both resilience and robustness. Highly centralized networks, where a few critical nodes handle a large proportion of the traffic, are inherently less resilient than decentralized networks with distributed control. A single point of failure in a centralized network can cripple the entire system, while a decentralized network can often reroute traffic around failures. Consider the difference between a star network and a mesh network; the latter's robustness is immediately apparent. Therefore, understanding the implications of different topological structures is key to designing resilient and robust systems.

Redundancy, in many ways, is the cornerstone of robust network design. This means having multiple pathways or backup systems in place to handle failures. If one route is disrupted, another can take over seamlessly. This is crucial for maintaining essential functions during unexpected events. In telecommunications, for example, redundant links ensure continuous connectivity even if one link fails. This concept extends beyond physical infrastructure; it also applies to software and data, where backup systems and data replication provide essential protection against data loss and system crashes. Implementing redundancy might increase initial costs, but the long-term benefits in terms of stability and reliability far outweigh these costs.

The concept of modularity is another key aspect that underpins a network's resilience and robustness. A modular network is composed of independent modules that can function relatively autonomously. If one module fails, the other modules continue to operate, minimizing the overall impact. This is similar to how the human body functions with independent organ systems. Failure of one system doesn't necessarily cause systemic collapse. This principle translates well to complex systems like software programs, where modular design allows for easier maintenance, updates, and fault isolation. Furthermore, modularity enhances adaptability; modifying or upgrading individual modules is simpler than changing an entire monolithic system.

Adaptive systems are those that can dynamically adjust their behavior in response to changing conditions. They possess inherent resilience because they can react to and recover from disturbances. Think of a biological ecosystem that adapts to environmental changes or a traffic control system adjusting to congestion. Such systems leverage feedback mechanisms to monitor their performance and make necessary adjustments. Designing for adaptability requires careful consideration of the system's dynamics and the development of appropriate feedback control strategies. This might involve incorporating self-healing mechanisms, dynamic routing protocols, or other self-organizing mechanisms that ensure the system's continued operation despite unexpected challenges.

However, resilience and robustness are not just about technical design; they also involve human factors. Effective training and emergency response protocols are essential to minimize downtime and ensure quick recovery from unexpected events. Furthermore, clear communication and coordination are vital for efficient disaster response. Robust systems are only as robust as the people operating and maintaining them. Regular maintenance, proactive monitoring, and robust security measures are all essential aspects of operational resilience. Investment in skilled personnel and robust training programs is, thus, an important element in maintaining operational resilience over the long term.

In conclusion, building resilient and robust networks requires a holistic approach. It's about combining careful topological design with redundancy, modularity, and adaptability. It's about anticipating potential failures, developing effective response mechanisms, and ensuring that human factors are fully integrated into the design and operational strategies. Ultimately, the goal is not just to prevent failures, but to minimize their impact and ensure that the network continues to function effectively, even under stress. Creating truly resilient systems requires a comprehensive understanding of network dynamics, and a commitment to integrating multiple strategies.

6 Applications of Network Science

Analyzing brain networks through the lens of network science has yielded incredible insights into cognitive functions and neurological disorders. Mapping brain connectivity reveals intricate patterns of interactions between brain regions, providing a framework for understanding how information is processed and integrated. For example, studies using diffusion tensor imaging (DTI) and functional magnetic resonance imaging (fMRI) have identified specific network hubs crucial for cognitive control and memory consolidation. Disruptions in these networks are often observed in neurodegenerative diseases like Alzheimer's and Parkinson's, highlighting the potential of network-based biomarkers for early diagnosis and personalized treatment. Further research might lead to interventions that target specific connections or network modules, optimizing therapies for these debilitating conditions. The complexity of this intricate system offers endless potential for scientific exploration.

Social networks, ubiquitous in our modern world, provide fertile ground for network science applications. By analyzing connections between individuals, we can uncover hidden structures and patterns of influence, revealing community formation and information diffusion dynamics. The spread of misinformation, for example, can be modeled using network science techniques, allowing for the development of strategies to counter its propagation. Understanding the network properties that contribute to the rapid spread of viral content can empower us to design more resilient and informed social media landscapes. This includes understanding how to identify and combat echo chambers, and creating interventions to mitigate the spread of false narratives. Moreover, applying network analysis to social movements reveals how collective action emerges and spreads through a population.

Technological networks, like the internet and power grids, represent critically important infrastructures that demand careful study. Analyzing the topological structure of these networks can reveal vulnerabilities to attacks or failures. By identifying critical nodes and links, we can develop strategies to enhance network resilience, increasing robustness against cascading failures. The cascading failure in a power grid, for instance, can cause significant economic disruption and societal damage. Understanding the network properties that lead to such failures can inform better infrastructure design and maintenance strategies. The application of network science here has significant implications for public safety and security, safeguarding critical infrastructure. Furthermore, understanding network dynamics allows for optimized network design and management, increasing efficiency and decreasing costs.

The implications of network science extend far beyond these three specific examples. Its principles are applicable across many diverse fields, from ecological networks to financial markets. Understanding network structure and dynamics is crucial for tackling complex challenges across these sectors, ranging from disease spread to economic stability. The interconnectedness of modern society mandates a network-centric approach to problem-solving, emphasizing the importance of considering relationships and interactions when addressing complex systems. Network science has already made considerable contributions; however, future advancements and interdisciplinary collaborations promise even more profound impacts on our understanding and management of the networked world. As we continue to generate more data, the opportunities to leverage network science will only grow exponentially. The development of better data-handling techniques and robust theoretical frameworks will accelerate these advances, providing innovative solutions to some of the world's most pressing problems. The potential applications of this field are vast and far-reaching.

6.1 Brain Networks

The human brain, a marvel of biological engineering, isn't just a collection of cells; it's a complex network, a vast interconnected system of neurons communicating through intricate pathways. Understanding this network architecture is crucial to unraveling the mysteries of cognition, consciousness, and neurological disorders. Mapping these connections, however, presents a formidable challenge. Researchers employ various techniques, such as diffusion tensor imaging (DTI) and functional magnetic resonance imaging (fMRI), to visualize these pathways and understand how different brain regions interact.

This intricate web of connections isn't static; its dynamism is a key characteristic. Brain networks are constantly adapting and reorganizing, reflecting our ever-changing experiences and internal states. Think of it as a living, breathing entity, constantly evolving in response to stimuli, learning processes, and even just the passage of time. This plasticity is what allows us to learn, adapt, and recover from injury. This dynamic interplay is far from random; the changes follow specific patterns and rules, which are still under investigation. The field of network neuroscience is dedicated to unraveling the principles governing these processes.

Identifying specific network hubs and their functions within the brain is another area of intense research. Some regions, like the default mode network (DMN), are highly active even during rest, suggesting a role in self-referential thought and introspection. Others, like the salience network, are involved in prioritizing sensory information and directing attention. Understanding how these hubs interact and coordinate their activity is paramount to comprehending higher-level cognitive functions. We are only beginning to scratch the surface of understanding these intricate relationships. Disruptions in these connections, as seen in neurological conditions, offer significant insight into the network's role in overall brain health.

The study of brain networks extends beyond simply identifying connections; it's about understanding the information flow within this complex system. How efficiently does information travel between different regions? Are there bottlenecks or delays that might impact cognitive performance? These are crucial questions for neuroscientists. The efficiency and robustness of these information pathways are believed to be significant factors in both healthy cognitive function and in neurological disorders. Research actively investigates methods to measure and quantify information flow, aiming to create more comprehensive models of brain activity.

Furthermore, the study of brain networks isn't confined to a single scale. Researchers investigate connections at various levels, from individual synapses to large-scale brain regions. This multi-scale approach is essential to gain a holistic understanding of brain function. We need to consider the interactions between genes, proteins, cells, and regions to understand the emergent properties of the whole brain. A unified model incorporating these different scales remains a grand challenge for the field. This complex interplay highlights the necessity of interdisciplinary collaboration to unravel the complete picture of the brain's intricate network architecture.

Delving deeper into the intricacies of brain networks reveals fascinating insights into both normal brain function and the pathogenesis of neurological disorders. Many neurological conditions, such as Alzheimer's disease and schizophrenia, are characterized by disruptions in network connectivity. Understanding these disruptions is key to developing effective treatments. For instance, analyzing changes in network topology, identifying weakened or severed connections, and observing changes in the timing and coordination of brain activity provides crucial clues about the underlying mechanisms of disease. These discoveries could lead to improved diagnostic tools and targeted therapies.

Beyond clinical applications, the study of brain networks opens exciting avenues for understanding the fundamental aspects of human cognition. By analyzing how information flows through the brain, and how different regions cooperate and compete for resources, we can gain a deeper understanding of our thoughts, feelings, and actions. This knowledge could contribute significantly to fields such as education, artificial intelligence, and even the development of brain-computer interfaces. The potential for transformative advancements in these areas is significant, underscoring the importance of continued research. These discoveries highlight the potential for creating more targeted educational strategies that capitalize on the brain's inherent plasticity and learning mechanisms.

In conclusion, the study of brain networks stands at the forefront of neuroscience research. By combining advanced imaging techniques with sophisticated analytical methods, we are steadily unraveling the mysteries of this incredibly complex system. The implications of this research are far-reaching, with potential benefits for understanding neurological disorders, enhancing cognitive performance, and deepening our understanding of the human mind. The journey of understanding the brain's intricate network is a continuous one, full of exciting discoveries and potential breakthroughs. As we delve deeper into this fascinating world, the possibilities seem endless. The future promises further advancements, unveiling even more of the brain's intricate secrets and pushing the boundaries of human knowledge. Our continued exploration into this realm is paramount not only for improving human health but also for unlocking the profound mysteries of consciousness itself.

6.2 Social Networks

Social networks are ubiquitous, influencing everything from our personal relationships to global political movements. Understanding their structure and dynamics is crucial, not just for sociologists, but for anyone navigating the modern world. We interact within these networks daily, exchanging information, building connections, and forming opinions. These actions, seemingly individual, contribute to a larger, complex system exhibiting emergent properties far beyond the sum of its parts. Think of the rapid spread of a viral trend – that's a testament to the power of network structure. The sheer scale and complexity of these networks often defy simple analysis. Consider, for instance, the intricate web of relationships on a platform like Facebook, a network with billions of nodes and trillions of connections. Mapping and analyzing such networks requires sophisticated tools and methodologies, often drawn from graph theory and statistical physics. Traditional social science methods struggle to capture the dynamism and interconnectedness inherent in these digital spaces. We need more than just surveys; we need to examine the flow of information, the patterns of interaction, and the evolution of the network itself.

One key aspect to consider is network topology. Are these networks primarily clustered, with dense groups of interconnected individuals, or are they more decentralized, with connections scattered randomly? The answer significantly impacts how information spreads and how influence is exerted. Understanding this topology is crucial for designing effective strategies for communication, marketing, or even disease prevention. In a clustered network, for example, targeting key individuals within tightly knit groups can be extremely effective; however, in a decentralized network, a broader, more widespread approach may be necessary.

The concept of centrality also plays a vital role in understanding social networks. Who are the key players? Who are the influencers, those individuals whose actions and opinions carry disproportionate weight? Identifying these central nodes is crucial for understanding how information disseminates and how influence is exerted. Centrality measures, like degree centrality (number of connections) or betweenness centrality (number of shortest paths passing through a node), help to pinpoint these influential individuals. Consider political movements, how one well-connected person can radically transform an entire community.

But it's not just about static structure. Social networks are constantly evolving. New connections are formed, old ones fade, and the flow of information shifts relentlessly. This dynamic nature demands the use of time-series analysis and agent-based modeling to fully appreciate the network's behavior. These sophisticated computational approaches allow researchers to simulate network evolution and test the impact of various interventions. For instance, simulating the spread of misinformation on a social network can reveal the vulnerabilities of the system and help strategize the prevention of harmful disinformation campaigns. This dynamic interplay of structure and flow is the heart of understanding these complex systems.

Beyond the theoretical frameworks, applying these concepts necessitates practical considerations. Data privacy is a major concern when studying social networks, especially in the era of big data. Ethically sourced, anonymized data, combined with robust methodological safeguards, are critical for maintaining privacy while providing valuable insights. This is more than just a technical hurdle; it's a foundational ethical responsibility when conducting research in this area. Without ethical considerations, the results, no matter how detailed, lose their scientific validity.

Moreover, the insights gained from studying social networks are not just academically interesting; they hold immense practical value. From marketing strategies to public health campaigns, understanding social network dynamics can inform more effective interventions. The spread of information is a perfect case in point; by identifying and understanding the propagation mechanisms, more targeted strategies can enhance the effectiveness of public health communications during disease outbreaks or vaccine awareness programs. The field offers a vast array of opportunities for impact beyond the classroom, moving the discussion from purely theoretical to directly applied knowledge.

Ultimately, the study of social networks presents a challenging yet rewarding avenue for research. The ever-evolving nature of these networks demands constant adaptation of methodologies and the development of novel techniques. The complexity is undeniable, but the potential for understanding human behavior, influencing social change, and improving our understanding of ourselves is immense. As we continue to explore these intricate patterns of connection, we can only hope to uncover more insights into this fascinating facet of human existence. We're merely scratching the surface, and what we discover holds vast implications for society. The ongoing research into these complex social formations will only enhance our capacity to navigate and improve upon the human condition.

6.3 Technological Networks

Technological networks are everywhere, shaping our daily lives in profound ways. From the intricate web of the internet connecting billions of devices to the complex infrastructure of power grids distributing energy across continents, these networks are critical to modern society. Their sheer scale and complexity pose unique challenges, demanding a sophisticated understanding of their structure, function, and vulnerabilities. Understanding these systems is paramount, given their impact on everything from global communication to national security. We must analyze their intricacies to ensure their continued reliable operation and adapt to their ever-evolving nature.

The internet, arguably the most prominent technological network, operates on a decentralized architecture. This distributed structure, while offering resilience against single points of failure, also introduces significant challenges in terms of security and control. The sheer volume of data flowing through this network necessitates robust security protocols to protect against cyberattacks and ensure data privacy. Maintaining the stability and integrity of the internet requires constant monitoring, proactive maintenance, and ongoing adaptation to emerging threats and technological advancements. This underscores the critical need for a deeper understanding of the underlying principles governing its operation.

Power grids, in contrast, often operate on a more centralized model. This can result in greater efficiency in some respects, but also introduces vulnerabilities. A single point of failure, such as a damaged transformer or a natural disaster, can cascade across the entire system, leading to widespread power outages. The increasing integration of renewable energy sources, such as solar and wind power, adds another layer of complexity. These sources are inherently intermittent, necessitating sophisticated control mechanisms to maintain the balance between supply and demand, and to prevent disruptions to the flow of electricity. Robust and adaptive control systems are vital for the seamless integration of these new energy sources and for ensuring the reliable delivery of electricity.

Transportation networks, including roads, railways, and air traffic control systems, represent another crucial area of technological networks. Efficient management of these networks is essential for the smooth flow of goods and people. However, unforeseen events like traffic accidents or severe weather can cause significant disruptions. Developing effective strategies for mitigating these disruptions, optimizing traffic flow, and improving overall network resilience is a continuous challenge. Advanced technologies, such as smart traffic management systems and predictive analytics, are playing an increasingly important role in improving the efficiency and reliability of these critical networks.

Communication networks, encompassing telephone lines, cellular networks, and satellite communication systems, are integral to modern communication. Their ability to efficiently transmit information over long distances is essential for various aspects of modern life, from personal communication to global financial transactions. However, the increasing demand for bandwidth and the ever-evolving nature of communication technologies necessitate constant upgrades and innovation. Moreover, ensuring the security and privacy of information transmitted across these networks is paramount, demanding sophisticated encryption techniques and proactive security measures. Understanding the vulnerabilities and limitations of these networks is critical for securing their continued reliable and secure operation.

These examples illustrate the diversity and complexity of technological networks. They highlight the need for a multidisciplinary approach to understanding these systems, drawing upon expertise in computer science, engineering, mathematics, and social science. By combining theoretical modeling with empirical analysis, we can gain valuable insights into the behavior of these networks, enabling us to design more robust, efficient, and resilient systems. This comprehensive understanding is vital for developing effective strategies to address the challenges presented by the complexity and criticality of these technologies, to maintain the uninterrupted flow of data, energy, transportation, and communication that underpins the functioning of modern society. Further research is crucial to navigate the challenges posed by rapid technological advancements. The future of these systems depends on this ongoing investigation and innovation. Failing to do so could lead to significant disruptions and vulnerabilities.

7 Future Directions and Open Questions

The inherent complexity of nonlinear networks presents significant challenges for future research. While we've made strides in understanding basic network structures and dynamics, predicting emergent behavior in highly interconnected, nonlinear systems remains a formidable task. Developing more sophisticated mathematical frameworks, capable of handling the inherent nonlinearities and high dimensionality, is crucial. This involves exploring advanced mathematical techniques beyond traditional linear analysis, potentially incorporating concepts from topological data analysis or information geometry to better capture the intricate relationships within these networks. Furthermore, the development of more powerful computational tools, capable of simulating and analyzing large-scale nonlinear networks, is essential to further progress. A key area needing further investigation is the role of noise and fluctuations in shaping network behavior. While deterministic models provide a foundation for understanding network dynamics, real-world networks are invariably subject to stochastic effects. A deeper understanding of how noise influences synchronization, bifurcation points, and overall network stability is vital, demanding the development of more robust stochastic models. Incorporating these stochastic elements could lead to improved predictive capabilities and a more realistic understanding of network resilience.

This requires careful consideration of the sources of noise, their statistical properties, and their interaction with the network's inherent nonlinearities.

The integration of network science with other fields presents exciting opportunities. For example, combining network science with machine learning techniques offers promising avenues for developing more accurate predictive models for complex network behavior. Machine learning algorithms could be used to identify hidden patterns and structures within large datasets of network data, leading to a deeper understanding of emergent phenomena and the development of more sophisticated control strategies. Specifically, deep learning architectures may offer a novel pathway to unravel the complexities inherent in nonlinear networks, allowing for the analysis of high-dimensional datasets that are currently intractable with conventional methods. This collaborative approach has the potential to accelerate discoveries significantly.

Furthermore, there is a critical need for improved methods for network visualization and analysis. As networks grow in size and complexity, the ability to effectively visualize and analyze network data becomes increasingly challenging. New visualization techniques, capable of revealing hidden structures and patterns in large-scale networks, are crucial for both researchers and practitioners. This includes exploration of 3-D visualization tools and methods for representing higher-order network interactions. Developing intuitive and interactive interfaces for exploring these visualizations is equally essential, allowing for more comprehensive analysis and intuitive discovery of patterns and relationships.

Beyond theoretical advancements, applications of network science are rapidly expanding, calling for innovative applications in diverse fields. For instance, understanding the dynamics of brain networks holds immense potential for advancing our understanding of neurological disorders and developing more effective treatments. Applying network principles to optimize the efficiency and resilience of technological infrastructure, such as power grids and communication networks, could have significant societal benefits, especially in the face of growing environmental stressors and increased reliance on complex interconnected systems. Finally, analyzing social networks can lead to a better comprehension of social dynamics and the spread of information. The ethical implications of these applications must be carefully considered, especially when dealing with sensitive data and potential biases within network structures. Finally, developing a deeper understanding of the relationship between network structure and function is a continuing frontier. While we've explored various network topologies and their influence on information propagation, much remains unknown about the precise mechanisms that give rise to specific functionalities within complex networks. Further investigation into the interplay between network structure and function will be crucial for designing robust and efficient networks across a range of applications. This includes a more comprehensive study of network motifs, their evolutionary origins, and their role in shaping network dynamics and functionality. A holistic approach, integrating various techniques from mathematics, computer science, and domain-specific expertise, is crucial for future advancements in this ever-evolving field.

7.1 Challenges and Opportunities

The field of network science, while burgeoning with exciting discoveries, faces significant hurdles. Developing robust analytical methods for exceptionally large and complex networks remains a substantial challenge. Current techniques often struggle to capture the intricate interplay of numerous nodes and connections, leading to incomplete or inaccurate models. This limitation hinders our understanding of emergent phenomena and limits the predictive power of our models, especially in dynamically changing systems like social networks or the rapidly evolving internet. Addressing this requires innovative computational approaches, perhaps incorporating machine learning techniques to handle the sheer volume of data and identify hidden patterns.

One significant opportunity lies in the integration of network science with other disciplines. Cross-disciplinary collaborations, particularly with biology, physics, and computer science, hold immense potential. Biologists could provide insights into the robust adaptive mechanisms observed in biological networks, offering blueprints for developing more resilient technological systems. Physicists could contribute their expertise in nonlinear dynamics, helping to unravel the complexities of chaotic behavior in networked systems. Computer scientists, meanwhile, can provide the tools needed to analyze and model these intricate structures, creating software tailored to the specific needs of network analysis.

The quest to understand and predict cascading failures in complex networks is paramount. The fragility of interconnected systems, highlighted by disruptions in power grids or financial markets, underscores the urgency of this pursuit. Improved understanding of how failure propagates through a network can guide the development of preventative measures and robust mitigation strategies. This necessitates the development of sophisticated models capable of simulating cascading failures, incorporating factors such as node vulnerability and the strength of interconnections. Such models could then be utilized to identify critical points of weakness and guide network design for enhanced resilience.

Another pressing challenge is the development of more sophisticated tools for visualizing and interpreting network data. The sheer volume of information inherent in large networks makes comprehension difficult. Currently available visualization methods, while helpful, often fall short when dealing with networks of immense size and complexity. Novel visualization techniques, possibly employing immersive virtual reality or augmented reality technologies, could prove invaluable. These methods would enable researchers to navigate and explore these vast networks more effectively, leading to deeper insights into their structure and function.

However, opportunities abound. The potential for impactful applications across various sectors remains immense. Network science offers the potential to revolutionize fields such as urban planning, disease control, and transportation management. The insights gained through analyzing network structures could enable optimized resource allocation, more effective strategies for disease prevention, and more efficient transportation systems. The key lies in tailoring network science principles to the specific contexts of these applications, incorporating relevant data and creating customized models.

Finally, we must consider the ethical implications of network science. The ability to analyze and predict behavior within social networks raises concerns about privacy and potential for manipulation. Therefore, responsible development and deployment of network science tools require careful consideration of ethical concerns, ensuring that the knowledge we gain is utilized for the betterment of society and not for malicious purposes. Establishing robust ethical guidelines and promoting open and transparent research practices are critical for navigating this landscape responsibly. The field's progress hinges on striking a balance between innovation and responsible application.

7.2 Unanswered Questions

The intricate dance of network dynamics, while extensively studied, leaves several key questions unanswered. Consider, for instance, the precise mechanisms driving the emergence of scale-free topologies in real-world networks. While preferential attachment models offer a plausible explanation, they often fail to fully capture the nuanced details observed in empirical data. Further investigation is needed to reconcile theoretical models with the observed heterogeneity and evolution of network structure, considering factors such as network growth dynamics, node-specific properties, and environmental influences. This requires not simply refined mathematical models, but also innovative experimental and data-driven approaches to fully grasp the complexity involved. The current limitations in our understanding hinder our ability to accurately predict and manage network behavior.

Understanding the interplay between network topology and dynamic processes remains a significant challenge. While research has established links between structural properties and various dynamical phenomena, a comprehensive theoretical framework that seamlessly integrates both aspects is still lacking. For example, the effects of network heterogeneity on synchronization, cascading failures, and information propagation are not completely understood. The current state of knowledge relies heavily on approximations and simplified models, often neglecting crucial details like node heterogeneity, edge dynamics, and time-varying network properties. This necessitates developing sophisticated mathematical and computational tools to accurately analyze the interplay between structure and dynamics in complex networks. Moving forward, interdisciplinary collaborations between mathematicians, physicists, computer scientists and domain experts will prove essential.

The robustness and resilience of real-world networks are often surprising, defying predictions based solely on average network properties. Networks, despite widespread interconnectivity and vulnerability, tend to exhibit a remarkable ability to withstand significant disruptions. The mechanisms behind this resilience are poorly understood, particularly in the face of targeted attacks or cascading failures. Improved understanding necessitates more sophisticated models that incorporate elements such as feedback loops, adaptive responses, and the interplay of multiple failure modes. Such models must also account for the diversity and heterogeneity of network components. A more granular approach is needed which focuses not only on the aggregate network behavior but also the behavior of individual components and subnetworks.

Predicting emergent behavior in large-scale networks remains a formidable challenge. While simplified models can capture certain aspects of collective behavior, they often fail to accurately predict the specific patterns that arise in realistic scenarios. The inherent complexity and high dimensionality of these systems make analytical solutions difficult if not impossible to find. Consequently, reliable prediction necessitates the development of robust and scalable computational tools, capable of simulating the behavior of intricate networks with many nodes and edges. Furthermore, novel approaches that exploit machine learning techniques, leveraging the power of data-driven methods, could greatly enhance predictive capabilities. Data from real-world networks should be used to train and validate the performance of these novel predictive approaches. Finally, the application of network science to real-world problems often encounters obstacles related to data limitations and model validity. Many real-world networks are poorly characterized, lacking complete data on their structure and dynamics. Furthermore, existing models often rely on simplifying assumptions which may not hold in specific contexts. Addressing these limitations requires the development of more robust methods for network data collection and analysis. It necessitates further research into developing and verifying more realistic models that capture the essence of complex systems' properties without neglecting their inherent intricacies. Ultimately, only through rigorous experimental validation and comparison with real-world observations can we advance our understanding of complex networks and their applications.

7.3 Future Research

Investigating the interplay between network topology and emergent dynamics remains a crucial area for future research. Specifically, we need to develop more sophisticated models that can accurately capture the complex interactions between network structure and nonlinear processes, going beyond simplistic assumptions often employed in current models. This includes a deeper understanding of how different topological features, such as modularity, community structure, and hierarchical organization, influence the onset and characteristics of phenomena like synchronization, cascading failures, and self-organized criticality. For example, we could explore the effects of network heterogeneity on the stability of synchronized states, examining how variations in node degree or connection strength impact the robustness of collective oscillations. Furthermore, understanding the impact of dynamic rewiring processes – where connections between nodes change over time – on emergent behavior is critical, requiring the development of analytical and computational techniques to handle time-dependent network topologies. This will help to better model real-world networks where connectivity is rarely static. Finally, we need to extend existing models to include a wider range of nonlinear elements, moving beyond simple oscillators and incorporating more realistic descriptions of individual node behavior, such as those reflecting stochasticity or adaptive responses.

The development of novel analytical techniques is essential to make progress in network science. Current methodologies often struggle to handle the complexity of large-scale networks exhibiting nonlinear dynamics. We urgently require more powerful mathematical frameworks that can efficiently analyze network structure, predict emergent behavior, and quantify the robustness of these systems. This includes developing more sophisticated methods for characterizing network topology beyond traditional measures like degree distribution and clustering coefficient. These new measures should capture the higher-order interactions that play a crucial role in determining network functionality and response to perturbations. Moreover, advanced computational algorithms are needed to simulate the behavior of extremely large and complex networks, incorporating realistic nonlinear interactions and noise. This computational power would facilitate the exploration of a wider range of parameter spaces and boundary conditions, allowing for a more comprehensive understanding of the emergence of complex behaviors. Combining these advanced analytical and computational approaches with techniques from information theory could offer a novel lens through which to examine the role of information flow in shaping network dynamics.

A particularly promising avenue for future research lies in the integration of network science with other disciplines. The power of network science lies in its universality and ability to provide a unifying framework for understanding diverse complex systems. Therefore, collaborative efforts across traditionally separate fields are crucial. By incorporating insights from fields such as statistical physics, dynamical systems theory, and control theory, we can gain a deeper understanding of the fundamental mechanisms underlying complex network behavior and explore more sophisticated control strategies. For example, collaborating with biologists could provide rich insights into the function of biological networks, such as the brain and immune system, enabling the development of more accurate and predictive models. Furthermore, collaboration with engineers could lead to the development of innovative technologies inspired by the principles of complex network design, potentially leading to improvements in areas like infrastructure resilience, communication networks, and the development of novel materials. A particularly interesting prospect lies in applying network science techniques to understand the emergent behavior of human societies, offering valuable insights for policymakers and social scientists aiming to address critical challenges like epidemics, misinformation spread, and socioeconomic inequality.

Ultimately, the most important aspect of future research in network science lies in asking the right questions. We need to move beyond the descriptive phase of simply characterizing network structure and dynamics, and focus more on developing predictive and prescriptive models. This necessitates a shift toward formulating more precise hypotheses, which can then be rigorously tested through carefully designed experiments and simulations. Furthermore, it is crucial to embrace interdisciplinary collaboration, fostering a broader range of perspectives and expertise, to accelerate the pace of discovery and innovation. This involves creating more opportunities for researchers from different fields to interact, fostering the development of shared language and methodologies. The ultimate goal is to translate our growing understanding of network science into tangible applications that can improve the lives of individuals and the wider society. By strategically focusing on critical questions, fostering cross-disciplinary collaboration and continuously pushing the boundaries of our analytical and computational tools, the next decade holds immense potential to unlock the deepest secrets of complex network systems.

7.4 Novel Applications

The intricate dance of neurons firing, echoing through the brain's complex architecture – this isn't just a biological marvel; it's a blueprint. Consider the potential of translating this understanding into revolutionary advancements in prosthetics. Imagine a prosthetic limb controlled not by clumsy external sensors, but by a direct neural interface, seamlessly integrated with the user's nervous system. This intricate connection wouldn't simply mimic movement; it would recreate the sensation of touch, the subtle nuances of pressure, the delicate feel of holding a fragile object. Such a prosthetic would be more than a replacement; it would be an extension of the self, restoring lost function and agency with an unprecedented level of precision. This is not science fiction. Researchers are actively exploring brain-computer interfaces that could make this a reality.

Furthermore, extending beyond prosthetics, consider the potential in treating neurological disorders. Parkinson's disease, for instance, is characterized by the disruption of neural pathways, leading to tremors and other debilitating symptoms. By meticulously mapping these pathways and employing targeted neuromodulation techniques, potentially through the use of advanced network analysis and targeted intervention within the neuronal network, the disease's impact might be significantly mitigated, restoring motor control and significantly improving the lives of affected individuals. This would necessitate a deep understanding of the network's dynamical behavior.

The application of network science is not limited to biological systems. Traffic flow, a seemingly chaotic system, offers another promising area. Intelligent traffic management systems could use network analysis to optimize traffic patterns, reducing congestion and improving commute times for millions. Real-time data from traffic sensors would be processed, enabling algorithms to dynamically adjust traffic signals, reroute traffic around bottlenecks, and ultimately, enhance efficiency. It's no longer just about reacting to congestion, but proactively shaping traffic flow with a precision never before possible. This would significantly alleviate urban gridlock and reduce wasted fuel.

Moreover, the burgeoning field of personalized medicine is poised to benefit greatly from these advances. Our understanding of complex biological networks – the intricate interplay of genes, proteins, and metabolic pathways – is continuously evolving. By leveraging network science techniques, we can better understand disease mechanisms, identify potential drug targets with greater accuracy, and even predict individual responses to treatments. This level of personalization will revolutionize healthcare, moving us towards more effective and safer therapies tailored to each patient's unique biological makeup. Such tailored treatment is a significant step beyond generalized medicine.

Beyond healthcare, we can consider the application to large-scale infrastructure. Power grids, for example, are complex networks that are vulnerable to failures. Applying our understanding of network resilience can lead to the development of more robust and self-healing systems. By analyzing the interconnectedness of power stations and transmission lines, we can identify vulnerabilities and enhance the grid's ability to withstand disruptions, preventing widespread blackouts and ensuring reliable energy distribution. Predictive modeling, based on network science, is crucial in developing such a system.

Finally, consider the world of finance. The global financial system is a complex network of institutions, markets and transactions. Network analysis could help identify systemic risks, predict financial crises, and enhance regulatory oversight. By understanding the interconnectedness of financial institutions, regulators can better assess the potential for cascading failures and take proactive measures to mitigate those risks. This isn't simply about reacting to crises; it is about developing a predictive framework, capable of anticipating the ripple effects of financial events, thereby increasing the stability of the global financial system. The integration of network theory with financial modeling has immense potential. These are but a few examples of the transformative potential of network science. The interconnectedness of seemingly disparate systems opens up an entire world of possibilities. Each breakthrough has implications far-reaching and profound.

7.5 New Perspectives

The exploration of complex networks often overlooks a crucial aspect: the inherent limitations of our current modeling approaches. We build elegant mathematical frameworks, but these are, by definition, simplifications. They often fail to capture the full richness and subtlety of real-world systems. For instance, consider the seemingly straightforward concept of network density. While we can calculate it precisely, the underlying dynamics influencing connectivity – the individual decisions, environmental pressures, and stochastic events – remain largely unmodeled, resulting in a gap between our models and reality. This suggests a need for dynamic, agent-based models which capture the decision-making processes of individual nodes. These would allow us to move beyond static representations and delve into the emergent properties arising from the interactions of adaptive units.

A promising area for future research lies in incorporating temporal dynamics more explicitly into our analyses. Many networks are not static entities; their structure and interactions evolve continuously. Understanding how these temporal changes affect network behavior, such as information propagation or resilience, is crucial. This necessitates the development of new analytical techniques, moving beyond static graph theory to encompass dynamic graph analysis. For example, we might employ time-series analysis to examine the evolution of network properties over time, allowing us to identify patterns and critical transition points that traditional static analyses miss. We could investigate how networks adapt to external shocks, such as sudden increases in traffic or catastrophic failures. Beyond improved modeling, we need new methods for analyzing massive datasets, the kind generated by modern networked systems. The sheer volume of data presents a computational challenge, requiring the development of more efficient algorithms and data structures. The current methods, efficient as they may seem, can struggle with the increasing scale and complexity of real-world network data. Moreover, much of the data is often incomplete, noisy, and inconsistent. Therefore, advanced data processing techniques and robust algorithms capable of handling this imperfect data are essential for extracting meaningful insights. Consider developing new machine learning techniques to identify subtle patterns and hidden structures that may be missed by traditional statistical methods. This may uncover emergent phenomena not previously recognized.

Moving towards a more holistic understanding necessitates interdisciplinary collaboration. Network science, as a field, thrives on the interplay of various disciplines. From mathematicians providing the theoretical frameworks to computer scientists supplying the computational tools and biologists offering insights into natural networks, each perspective adds to our collective knowledge. But there are still untapped avenues of collaboration. Incorporating expertise from areas like economics, sociology, and psychology will provide further richness to the models and offer a more nuanced view of human decision-making within network structures. In examining social networks, for example, a psychological understanding of decision-making processes would enrich the modeling of information diffusion and the spread of influence.

Finally, we must consider the ethical implications of our research. As we gain a deeper understanding of complex networks, we gain the power to influence their behavior. This power brings responsibility. Consider, for instance, the potential to manipulate social networks for political purposes or to exploit vulnerabilities in technological infrastructures. It is critical that we, as researchers, carefully consider the ethical ramifications of our work and strive to use our knowledge responsibly. Therefore, alongside advancements in methodology and modeling, a strong ethical framework needs to be established to guide the use of network science and its applications. The future of network science lies not just in its continued technological advancements, but in its responsible application.

www.ingramcontent.com/pod-product-compliance
Lightning Source LLC
LaVergne TN
LVHW022022200926
848481LV00007B/2492
9798349217043